Learn 2 Think

200 Challenging Math Problems

Every 3rd Grader Should Know

If you have enjoyed problem solving with this book, we would be happy to receive your reviews on Amazon.com.

This book belongs to:

_____

_____

200 Challenging Math Problems

every 3rd grader should know

New edition 2018
Copyright Learn 2 Think Pte. Ltd.

Published by:
Learn 2 Think Pte. Ltd.

ISBN: 978-981-07-2764-2

Master Grade 3 Math Problems

Introduction:

Solving math problems is core to understanding math concepts. When Math problems are presented as real-life problems students get a chance to apply their Math knowledge and skills. Word problems progressively develop a student's ability to visualize and logically interpret Mathematical situations.

This book provides numerous opportunities to every student to practice their math skills and develop their confidence of being a lifelong problem solver. The multi-step problem solving exercises in the book involve several math concepts. Student will learn more from these problems solving exercises than doing ten worksheets on the same math concepts. The book is divided into 10 chapters. Within each chapter questions move from simple to advance word problems pertaining to the topic. The last chapter of the book explains step wise solutions to all the problems to reinforce learning and understanding.

How to use the book:

Here is a suggested plan that will help you to crack every problem in this book and outside.

Follow these 4 steps and all the Math problems will be a NO PROBLEM!

Read the problem carefully:

- What do I need to find out?
- What math operation is needed to solve the problem? For example addition, subtraction, multiplication, division etc.
- What clues and information do I have?
- What are the key words like sum, difference, product, perimeter, area, etc.?
- Which is the non-essential information?

Decide a plan

- Develop a plan based on the information that you have to solve the problem. Consider various strategies of problem solving:
- Drawing a model or picture
- Making a list
- Looking for pattern
- Working backwards
- Guessing and checking
- Using logical reasoning

Solve the problem:

Carry out the plan using the Math operation or formula you choose to find the answer.

Check your answer

- Check if the answer looks reasonable
- Work the problem again with the answer
- Remember the units of measure with the answer such as feet, inches, meter etc.

Master Grade 3 Math Problems

Note to the Teachers and Parents:

✎ Help students become great problem solvers by modelling a systematic approach to solve problems. Display the 'Four step plan of problem solving' for students to refer to while working independently or in groups.

✎ Emphasise on some key points:

✎ Enable students to enjoy the process of problem solving rather than being too focused on finding the answers.

✎ Provide opportunities to the students to think; explain and interpret the problem.

✎ Lead the student or the group to come up with the right strategy to solve the problem.

✎ Discuss the importance of showing steps of their work and checking their answers.

✎ Explore more than one possible solution to the problems.

✎ Give a chance to the students to present their work.

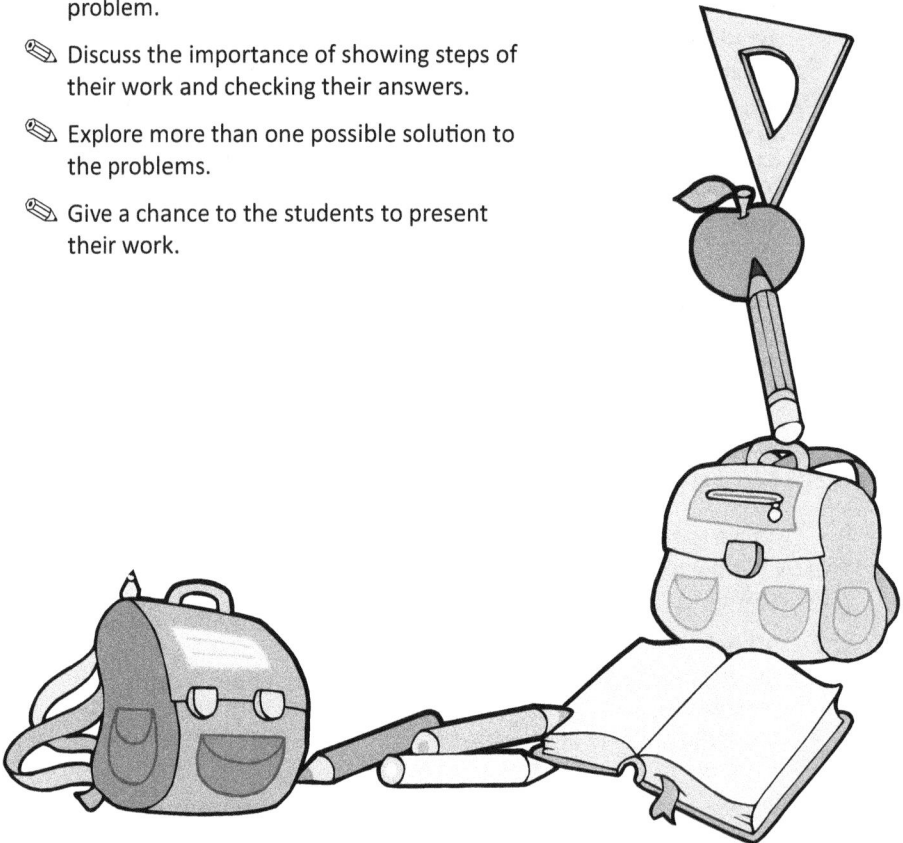

Contents

PROBLEM 1

I am a four digit number. I am smaller than 6000 but bigger than 5000. My hundreds digit is smaller than 8 but bigger than 6. My tens digit is an odd number, smaller than 5 but bigger than 1. My units digit is in the 3 times table and is bigger than 4 but smaller than 8. What number am I?

Answer: ....

PROBLEM 2

I am a three digit number. My units digit is 72 less than 81. The hundreds digit is an odd number which is bigger than 1 but smaller than 4. The tens digit is the same as 3 + 4. What number am I?

Answer: ....

PROBLEM 3

My number has five digits. The ten thousands digit is the same as the hundreds digit, and it is an odd number which can be divided by 3 to make 3. The thousands digit is one bigger than 8. The units digit is 2 more than the tens digit. The tens digit is smaller than 1. What is my number?

Answer: ....

My number has five digits. The thousands digit is the number of days in a week. The tens digit is the number of fingers on your hand. The units digit is half of 8. The hundreds digit is 4 more than the tens digit. The ten thousands digit is 1 + 2 + 3 + 6 − 7. What is my number?

Answer: ....

PROBLEM 5

How many times does the digit nine appear if you write down all the numbers from 1 to 100?

Answer: ....

Shown below are four number cards.

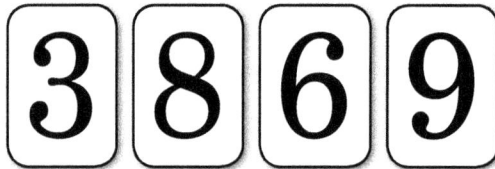

3 8 6 9

Use these numbers without repeating, and make the smallest 3-digit number.
Then, using these numbers without repeating, make the largest 2-digit number.
What is the difference between the smallest 3-digit number and the largest 2-digit number you made?

Answer: ………………………

Here are three digits.

1 5 8

Use these digits and make a number between 160 and 190.

Then, using these numbers, make a number between 550 and 600.

What is the sum of both the numbers that you made?

Answer: ....

PROBLEM 8

Here are three number cards.

9 5 7

a) Write down the smallest number you can make using the three cards.

b) Write down the number closest to 760 you can make using the three cards.

Answer: ....

PROBLEM 9

Write the number names given in each sentence in digits.

a) Last month four million three hundred thousand children watched American Idol.

b) High School Musical sold two hundred and thirty thousand five hundred and sixty copies of their DVD last year.

c) Taylor Swift's songs have been downloaded two hundred and twenty two thousand, six hundred and five times in this year.

Answer: ....

PROBLEM 10

Find the sum of the largest two-digit even number between 80 and 90 and the smallest two-digit even number between 20 and 30.

Answer: ....

PROBLEM 11

Make the largest 3-digit number with all the digits different and that is divisible by 5.

Answer: ....

PROBLEM 12

Find the smallest three-digit number that satisfies the following conditions:

1. All digits are different.
2. None of the digits is 0.
3. Every digit is divisible by 3.

Answer: ....

PROBLEM 13

If you take away one fourth of me and then add two, you get 17. What number am I?

Answer: ....

PROBLEM 14

What number am I?

a) I am four less than 21.

b) I am half of 26.

c) I am a quarter of 4.

d) I am a third of 9 added to 12.

e) If you add 20 to me you get 50.

Answer: ....

PROBLEM 15

Find a pair of numbers with:

a) a sum of 11 and a product of 24.

b) a sum of 40 and a product of 400.

c) a sum of 15 and a product of 54.

Answer: ....

In a street, houses are numbered with consecutive odd numbers from 1 to 21 on one side. On the other side of the street, the houses are numbered with the consecutive even numbers from 2 to 16. How many houses are there on the street?

Answer: ………………………

PROBLEM 17

The sum of the four consecutive odd numbers is 40. Find the largest of these 4 odd numbers.

Answer: ....

PROBLEM 18

One fountain pen costs as much as 2 ball point pens. Also, 1 ball point pen costs as much as 4 pencils. How many pencils can you buy for the cost of one fountain pen?

Answer: ....

PROBLEM 19

There were 15 more boys than girls in a class. 8 boys wore spectacles and 5 more girls than boys wore spectacles. If all the girls in the class wore spectacles, how many children were there in the class?

Answer: ....

Tony eats a 1/4 bag of popcorn every day. How many bags of popcorn would be needed to have enough popcorn for him to eat for two weeks?

Answer: ....

PROBLEM 21

The number 202 is the same when you write it in the reverse order. Such numbers are called palindromes. What is the sum of 202 and the next palindrome number bigger than 202?

Answer: ....

PROBLEM 22

Jenny is reading a story book. The sum of the two facing pages that she is reading is 37. What are the two facing page numbers?

Answer: ....

PROBLEM 23

Jim placed 12 cans in a row. He had fifty $1 coins in a bag.
He took one $1 coin from his bag and put it in the first can, then took two $1 coins from his bag and put in the second can, followed by three $1 coins that he put in the third can, and so on he continued.
After how many cans will he have 14 coins left in his bag?

Answer: ....

PROBLEM 24

Jason used some coins to make a triangle. He placed one coin on each corner of the triangle. There were 5 coins on each side of the triangle. How many coins did Mark use?

Answer: ....

PROBLEM 25

Jenny puts 9 coins in a row on a table. Each coin is 10 cm apart. How far is the ninth coin from the first one?

Answer: ....

PROBLEM 26

The houses in a street are numbered sequentially from 1 to 24. How many times does the digit 2 occur in the numbers of those houses?

Answer: ....

Mark had 61 matches. Using some of them, he made a triangle whose sides had 7 matches each. With the remaining he made a rectangle whose one side had 6 matches. How many matches long was the rectangle's other side?

Answer: ....

A clock cost $24 and a watch cost $15 more than the clock. Daniel bought 2 clocks and a watch. How much money did he spend?

Answer: ....

There were 206 men at a table tennis tournament. There were 49 fewer men than women.

a) How many women were there?

b) How many people were there altogether?

Answer: ....

PROBLEM 30

Jenny was going to Disneyland. She travelled 150 km by car, 50 km by bus and 3250 km by plane. How far is Disneyland from her house?

Answer: ....

PROBLEM 31

Joshua is 12 years old now. His father is 32 years older than him. What will be their total ages in 5 years time?

Answer: ....

Jenny has 230 oranges. Sarah has 145 more oranges than Jenny. Jason has 310 more oranges than Sarah. How many oranges do the three children have?

Answer: …………………………

PROBLEM 33

985 participants in a 5 km cross country race are females. The number of male participants is 1297 more than the number of female participants. How many participants are there altogether?

Answer: ....

PROBLEM **34**

Regina had 1050 magazines. Sophia had 672 fewer magazines than Regina. How many magazines did Sophia have?

Answer: ………………………

PROBLEM 35

Sarah and Ben together have 3540 stamps.
Sarah has 978 stamps. How many fewer
stamps does Sarah have than Ben?

Answer: ....

PROBLEM 36

There are 389 cookies in a box. If the box can hold 3160 cookies, how many more cookies can Shirley put in the box?

Answer: ....

PROBLEM 37

There are the same number of elephants and tigers in a circus. There are 20 less acrobats than clowns. If there are 200 circus members altogether and 60 are clowns, how many elephants, tigers, and acrobats are there?

Answer: ....

Cherry saved $190 in a week. She saved $48 on Monday and $24 on Tuesday. How much money did she save during the rest of the week?

Answer: ………………………..

PROBLEM 39

A bus leaves at 16:20 p.m. with 24 people traveling in it and reaches Harbor Centre at 17:52 p.m.17 people get off the bus and 8 get on at Harbor Centre. How many people will still be on the bus at 17.54?

Answer: ....

PROBLEM 40

The difference between two numbers is 150. If the smaller number is 2356, find the bigger number. Find the sum of the two numbers.

Answer: ....

PROBLEM 41

After Jeffrey gave 20 sweets to Adeline, she had 70 sweets and he had 10.

a) How many sweets did Jeffrey have at first?

b) How many sweets did Adeline have at first?

Answer: ....

PROBLEM 42

In a car park, there are 198 buses. There are 54 more buses than vans and 36 more motorcycles than vans. How many vehicles are there altogether?

Answer: ....

PROBLEM 43

How many fifty-cent coins can you exchange for $100?

Answer: ....

PROBLEM 44

Kevin has 7 times as many paper clips as
Richard. Richard has 349 paper clips.
How many paper clips does Kevin have?

Answer: ....

PROBLEM 45

Caroline baked 7 trays of cupcakes. There were 16 butter cupcakes and 10 chocolate cupcakes on each tray. How many more butter cupcakes than chocolate cupcakes did Caroline bake?

Answer: ………………………..

PROBLEM 46

Samuel sold 152 oranges on Monday and 62 oranges on Tuesday. He had twice as many oranges left as the total number of oranges sold. How many oranges did he have left?

Answer: ....

PROBLEM 47

There are 328 stamps in 1 album. How many stamps are there in 6 such albums?

Answer: ....

PROBLEM 48

There are 27 boys and 15 girls in a class. Each student has 8 exercise books. How many exercise books do they have altogether?

Answer: ....

PROBLEM 49

One bag can contain 168 sweets. What is the total number sweets if there are 7 bags?

Answer: ....

There are 110 black chairs and 4 times as many white chairs as black chairs in an office. The white chairs are arranged in rows of 8. How many rows of white chairs are there?

Answer: ………………………

There are 220 marbles in a box. 80 of them are red and the rest are green and yellow. If there are twice as many red marbles as green marbles, how many yellow marbles are there?

Answer: ....

PROBLEM 52

There are 5 times as many mangoes in box A than in box B. There are a total of 60 mangoes in both the boxes .What is the number of mangoes in each box?

Answer: ....

A box of 10 eggs cost $8. If the eggs are bought individually, they cost $2 each. How much more would it cost to buy 20 eggs individually than in boxes?

Answer: ....

PROBLEM 54

Dennis had 310 apples and 230 oranges. He sold twice as many apples as oranges and had 240 fruits left. How many apples did he sell?

Answer: ....

PROBLEM 55

A buffet lunch costs $23 for an adult and $7 less for a child. Find the total amount of money that a family of 2 adults and 3 children has to pay for the buffet lunch.

Answer: ....

PROBLEM 56

There are 110 green stickers and twice as many blue stickers as green stickers. The rest are red stickers. If there are 85 red stickers, how many stickers are there altogether?

Answer: ....

PROBLEM 57

Marcus has 120 red balloons and 210 green balloons. Jacquelyn has twice as many balloons as Marcus and Dennis has 3 times as many balloons as Marcus. How many balloons do they have altogether?

Answer: ....

PROBLEM 58

There are 120 apples and oranges in a box. The number of oranges in the box is twice the number of apples .

a) How many apples are there?

b) How many oranges are there?

Answer: ....

PROBLEM 59

Gavin arranged 25 rows of 7 chairs in a hall. 137 chairs were occupied. How many chairs were not occupied?

Answer: ....

PROBLEM 60

In a school basketball tournament, there were a total of 280 students, comprising of Americans, Chinese and Indians. 40 American students left the basket ball tournament and the remaining students were all Chinese and Indians. The number of Chinese students were 5 times more than the number of Indian students. How many more Chinese students than the Indian students were there at the tournament?

Answer: ....

PROBLEM 61

There are 175 cookies in 1 box. Margaret has 6 such boxes of cookies. After she sold 368 cookies to Kelvin, how many cookies did she have left?

Answer: ....

Jason bought 4 pairs of shoes at $58 each and 2 pairs of socks at $5 each. He then had $85 left. How much money did he have at first?

Answer: ………………………

PROBLEM 63

Mary has 95 books. David has thrice as many books as Mary. Rose has the same number of books as David.

a) How many books does David have?

b) How many books do the three children have altogether?

Answer: ....

PROBLEM 64

After giving 128 marbles to each of the 7 boys, Mrs Smith had 34 marbles left. How many marbles did the 7 boys receive altogether? How many marbles did Mrs Smith have at first?

Answer: ....

PROBLEM 65

After arranging 158 stamps into each of the 5 albums, Ahmad had 97 stamps left. How many stamps did he have at first?

Answer: ....

PROBLEM 66

Kelvin spent $360. Jane spent half as much as him. How much did they spend altogether?

Answer: ………………………

PROBLEM 67

104 children were going to the zoo in mini buses. Each mini bus could seat 8 children. What was the least number of mini buses needed to take all the children to the zoo?

Answer: ....

PROBLEM 68

Mrs. Thomas baked 104 apple pies and 216 lemon pies. She packed the two different types of pies separately into boxes of 4 each. How many more boxes of lemon pies than apple pies did she have?

Answer: ....

PROBLEM 69

Claire has 840 sweets. She has 4 times as many sweets as Caroline. How many sweets do they have altogether?

Answer: ....

There were 10 balloon in a packet. Caroline bought 8 such packets. She blew them and tied them into bunches of 5. How many bunches of balloons did she have?

Answer: ………………………

2 similar limited edition books cost $186. Doris bought 3 such books and had $28 left. How much money did she have at first?

Answer: ....

PROBLEM 72

Doris bought 8 packets of tomatoes to use in her cooking class. Her cooking class has 25 students. When she gave each of them 4 tomatoes, she had 4 tomatoes left. How many tomatoes were there in each packet that she bought?

Answer: ....

A shopkeeper sold 230 potatoes on Monday and twice as many potatoes on Tuesday and had 150 potatoes left. If he wanted to sell an equal number of potatoes within 2 days, how many potatoes should he sell on each day?

Answer: ....

PROBLEM 74

9 packets of sweets cost $72 and 3 packets of biscuits cost $12. Amie wants to buy 4 packets of sweets and 2 packets of biscuits. Find the total amount of money that she has to pay for these items.

Answer: ....

Florence bought 18 packets of apples weighing 5 kg each. She then packed them into plastic bags so that each plastic bag has 2 kg of apples. How many plastic bags did she use?

Answer: ....

PROBLEM 76

Henry had 542 carrots. He sold 138 carrots and packed the remaining carrots equally into 4 boxes. How many carrots were there in each box?

Answer: ....

PROBLEM 77

8 people paid $90 each for a gala dinner. The organizers were still short of $24. What amount should each of people pay so that the organizers are not short of money.

Answer: ....

PROBLEM 78

There were 250 fiction books and 180 non-fiction books in a library. There were 3 times as many non-fiction books as science books. How many books were there altogether?

Answer: ....

Tracy bought 2 similar watches and had $150 left. In total, she spent 3 times the amount of money left with her. How much did each watch cost?

Answer: ....

PROBLEM 80

Grade 3 is having a pizza party. The teacher ordered 6 cheese pizzas, 9 chicken pizzas, and 6 vegetarian pizzas. If each pizza is divided into 6 pieces and each student gets 3 pieces, how many students are there in Grade 3?

Answer: ....

There are 24 potatoes in a box. Thomas bought 2 boxes and used 12 potatoes. He then repacked the remaining potatoes into packets of 6. How many packets of potatoes does he have?

Answer: ....

PROBLEM 82

An adult ticket to the zoo safari cost $18.50. A child ticket cost $7.50. Florence took her three children to the zoo. After paying for all the tickets, she had $56 left. How much money did she have at first?

Answer: ………………………

PROBLEM 83

There are 8 charity tickets in a booklet. Linda sold 18 booklets and Jennifer sold 4 times as many booklets as her. Michael sold half the number of booklets that Linda sold. How many charity tickets did they sell altogether?

Answer: ....

Caroline had 6 times as many sweets as Melvin. They had a total of 28 sweets. How many sweets should Caroline give to Melvin so that both have the same number of sweets?

Answer: ....

Box A contains 250 marbles and Box B contains half as many marbles as box A. Box C contains 3 times as many marbles as box B. Find the total number of marbles in the 3 boxes.

Answer: ....

PROBLEM 86

A bottle of detergent cost $9 before the sale. During the sale, the price of the bottle of detergent was $6. Daniel bought 8 bottles of detergent before the sale. How many more bottles of detergent could he buy if he had spent the same amount of money and bought the detergent during the sale?

Answer: ....

Jessica packed 23 boxes of 6 cupcakes each. She then had 3 cupcakes left. How many more boxes would she need if she had instead packed all the cupcakes into boxes of 3 cupcakes each?

Answer: ....

PROBLEM 88

Betty has 149 buttons. She needs to put the buttons equally into 6 drawers. How many buttons will not get put into drawers?

Answer: ....

PROBLEM 89

If it takes 20 minutes to cut a log into 5 pieces, how long would it take to cut a similar sized log into 10 pieces?

Answer: ....

PROBLEM 90

Peter has 24 meters of fence and wants to make the biggest rectangular yard possible for his dog to run around. What length should he make each side of the yard?

Answer: ....

PROBLEM 91

All the 132 students of Grade 3 are going for a movie trip at the local cinema. Each rows of the movie theater has 11 seats. How many full rows at the theater will the students of Grade 3 use?

Answer: ....

PROBLEM 92

There 9 rows of 24 chairs in the school hall. Mr Tim rearranged them into rows of 6. How many rows of chairs did Mr Tim rearrange them into?

Answer: ....

There are 756 stamps in Jane's collection. If the stamps are organized equally into 6 stamp albums, how many stamps are there in each album?

Answer: ....

PROBLEM 94

Jerry had 4 boxes of crayons and each box had 24 crayons. He lost 8 crayons and his sister broke another 16 crayons. He put the remaining crayons into two new packets. How many crayons will Jerry put in each packet?

Answer: ....

PROBLEM 95

Alice went picking flowers and came back with three baskets full of flowers. Each full basket contained 66 flowers. Alice then made bunches of 12 flowers each and threw away the remaining How many flowers did Alice throw away?

Answer: ....

There were 28 blue plates and 29 green plates in Mrs Anderson's kitchen. There were also thrice the number of bowls as green plates. Find the total number of plates and bowls in the kitchen.

Answer: ……………………………

PROBLEM 97

A third of the class has pets. There are 45 students in the class. How many students have pets?

Answer: ....

PROBLEM 98

One fifth of the thirty five members in the drama club are girls. How many girls are there in the drama club?

Answer: ………………………

PROBLEM 99

A cake is cut into fifteen equal pieces. Sam eats five pieces at the break time. What fraction of the cake is left?

Answer: ....

PROBLEM 100

Two third students out of a class of 30 understand a math problem. What fraction of the class doesn't understand the math problem?

Answer: ....

PROBLEM 101

One quarter of the school walks to school.
Another quarter comes to school cycling.
What fraction of the school does neither?

Answer: ....

PROBLEM 102

One quarter of a box of chocolates has been eaten. If the box had 12 chocolates, what fraction of chocolates are left?

Answer: ....

PROBLEM 103

Grade 3 has 40 students. 1/4 of the students have i-pads and another one half have an i-pod Touch.

a) How many children have i-pads?

b) How many have an i-pod Touch?

c) How many of the class have neither?

Answer: ....

Siya has 40 cm of ribbon. She wants to cut the ribbon into equal pieces. Work out the different ways she can cut the ribbon with no bits left over.

What fraction of the whole ribbon is each part?

Size of each piece	No of pieces	Is this whole number of cms	Fraction of whole

Answer: ....

PROBLEM 105

Charles wants to put 15 ml of fruit punch equally into glasses. He wants an exact whole number of ml in each. Work out the different ways he can put the fruit punch. What fraction of the whole amount is in each glass?

No. of glasses	Amount in each glass	Fraction of the whole

Answer: ....

A pizza has been divided into 8 equal slices. Ben eats a quarter of the pizza and Tim eats another quarter. How many slices are still left?

Answer: ....

PROBLEM 107

Moshi monster cards are selling for a discounted price; they normally cost $4.50 per packet but have been reduced to half price. How much will a packet cost now?

Answer: ....

Peter had $80 in his wallet and he spent $60 out of it. What fraction of his money did he spend?

Answer: ....

PROBLEM 109

I spent $69 at a book sale, and also spent two third of that amount to buy a present for my friend. How much did I spend to buy the present?

Answer: ....

In a Grade 3 class there are 30 girls and 20 boys. What fraction of the class are girls?

Answer: ....

PROBLEM 111

If I eat 1/2 of a cake and you eat 2/4 of the cake, how much cake would be left?

Answer: ....

PROBLEM 112

A recipe for baking one cake uses 1/2 cup sugar, 1/2 cup brown sugar, 1/4 cup flour, and 4 eggs. If I want to bake 2 cakes how much of each ingredient would I need?

Answer: ....

PROBLEM 113

There are 8 pigs, 4 cows, and 6 roosters at a farm. What is the fraction of each animal at the farm?

Answer: ....

PROBLEM 114

There are 120 children performing in the school annual concert. Half of the students are given 4 invitations for their family to come to the concert. The other half is given 2 invitations each. How many people were invited in all to see the concert.

Answer: ....

PROBLEM 115

Tickets for Lady Gaga's concert are selling at $320 each. They are on sale today for half price. How much will a ticket on sale cost?

Answer: ....

PROBLEM 116

William earned $3500 a month. He spent $1420 on his children and $1200 on transport and food. He then saved the remaining money. At this rate, how much would William save in 3 months?

Answer: ....

PROBLEM 117

Samuel wanted to buy a shirt that cost $45 and a pair of jeans that cost $52. He was short by $13. How much money did Samuel have?

Answer: ....

PROBLEM 118

An orange cost $1.50 and a guava cost 50 cents more than the orange. Jessica bought an orange and a guava and had $15 left. How much did she have at first?

Answer: ....

Henry had $35. He bought 5 boxes of pizza and had $10 left. How much will 8 boxes of pizza cost?

Answer: ....

PROBLEM 120

Gavin earns $4350 every month. He spends $350 on transport and $120 more on food than transport. He then gives $600 to his wife and saves the rest. How much does he save?

Answer: ....

PROBLEM 121

7 books cost $105. Victor bought 9 such books and had $14 left. How much money did he have at first?

Answer: ....

PROBLEM 122

Richard spent $70.50 on a camera and $55.50 less on a radio. How much money did he spend on both the items?

Answer: ....

PROBLEM 123

A toaster cost $87. An oven cost $150 more than the toaster . A Juicer cost $28 less than the oven. Elizabeth bought the 3 electrical items and gave a $1000 note to the cashier. How much change will she get back?

Answer: ....

Florence packed 100 apples equally into boxes of 5 and sold each box for $3. Oliver packed 100 oranges equally into boxes of 4 and sold each box for $2. How much more did Florence earn than Oliver?

Answer: ....

PROBLEM 125

Samuel saved $564 each month. After 6 months, he spent some of his savings. If he had $1205 left after spending, how much did he spend?

Answer: ....

PROBLEM 126

An electronic dictionary cost $78. A cake mixer cost $22 more than the electronic dictionary and a toaster cost half as much as the cake mixer. What was the total cost of the 3 electrical items?

Answer: ....

PROBLEM 127

Maggie had $290 at first. When she gave $50 to Jeffrey, she still had thrice as much money as Jeffrey. How much money did Jeffrey have at first?

Answer: ....

PROBLEM 128

Kelvin took his wife and 3 children to a restaurant for dinner. He paid a total of $30.40. If each adult's dinner cost $8.00, how much did he pay for each child?

Answer: ....

William spent $213.50 on Saturday. He spent $89.00 more on Sunday than on Saturday. How much money did William spend on both the days combined?

Answer: ....

PROBLEM 130

At a book sale, Larry bought 9 English books and 3 Chinese books for a total of $330. Each Chinese book cost $35.00. How much did each English book cost?

Answer: ....

PROBLEM 131

Susan bought 6 rings at $48 each and gave the cashier three hundred-dollars.

a) How much did the 6 rings cost in total?

b) How much change did she receive from the cashier?

Answer: ....

PROBLEM 132

Melvin bought 4 clocks and a T-shirt for $320. Each clock cost $73.

a) How much did the T-shirt cost?

b) How much more did each clock cost than the T-shirt?

Answer: ....

PROBLEM 133

Lester bought 8 chairs and a table for $1085.
The table cost $389.

a) How much did each chair cost?

b) How much did Lester pay for one chair
   and a table put together?

Answer: ....

PROBLEM 134

The total cost of a pencil case and a pencil is $2.40 . The pencil case cost $1 more than the pencil. Find the cost of the pencil.

Answer: ....

PROBLEM 135

How many twenty five cent coins (quarters) can I exchange for $10?

Answer: ....

PROBLEM 136

Betty paid $16 for each of the 4 books she bought and had $10 left. How much money did she have at first?

Answer: ....

PROBLEM 137

Joe and Julia spent $13.75 each. If Joe, Julia and Bobby spent $60.00 altogether, how much money did Bobby spend?

Answer: ....

PROBLEM 138

Mr. Tan bought 9 books at $8 each and a ring for $68.50. How much did he spend altogether?

Answer: ....

A lamp costs $26. An oven costs $89 more than the lamp. What is the cost of the oven?

Answer: ....

PROBLEM 140

Laura had $230. She spent $120 on cosmetics and some of the remaining money on jewellery. She then had $60 left. How much did she spend on jewellery?

Answer: ....

PROBLEM 141

Henry saves $285 a day. How much money will he save in 9 days?

Answer: ....

Find the cost of 4.5 kilograms of sugar at 20 cents per 500 grams.

Answer: ....

PROBLEM 143

After spending $269 on Monday and $195 more on Tuesday than on Monday, Betty saved the rest of her money. She saved $1965.

a) How much did she spend altogether?

b) How much money did she have at first?

Answer: ....

PROBLEM 144

The total cost of a Math book and an English book is $27. The Math book costs $3 more than the English book. Find the cost of the Math book.

Answer: ....

PROBLEM 145

Kevin saves $45.80 a day. How much can he save in a week?

Answer: ....

A lorry costs 9 times as much as a motorcycle. If the motorcycle costs $1078, find the cost of the lorry.

Answer: ....

PROBLEM 147

A watch costs 4 times as much as a pen. If the pen costs $6.80;

Find the cost of the watch.

Find the total cost of the watch and the pen.

Answer: ....

6 boys paid $174 for a present and had $36 left.

a) How much did each boy pay for the present?

b) How many pens could they buy with the remaining amount of money if each pen cost $4?

Answer: ....

PROBLEM 149

Jake wants to watch a television program that will begin at 9:00 p.m. His mom says he can watch it but he must first do his chores and finish his homework. It takes him half an hour to do his chores and another half an hour to finish his homework. At what time should Jake begin if he is to finish in time to watch the program?

Answer: ....

PROBLEM 150

The solar eclipse began at 2:13 in the afternoon and ended at 3:51 p.m. How long did the eclipse last?

Answer: …………………………

PROBLEM 151

Write the correct time for each of the following:

a) Tina's alarm rings at 6 o'clock in the morning. Four and a half hours later she leaves the house. What time does she leave?

a) Tina and Ricky leave their home at 9 o'clock. They arrive at the park at 1:30 pm. How long do they take to walk to the park?

Answer: ....

PROBLEM 152

a) Ben and Mark reached an ice cream shop at 9.15 a.m. but they found out that the ice-cream shop does not open until 11.30 a.m. How long did they have to wait until the shop opened?

b) At 11.45 a.m. they walked to the library and that took them 20 minutes. They read books there for 30 minutes and then Mark looked at his watch. What time was it then?

Answer: ....

PROBLEM 153

A train takes 4 minutes to reach the next station. It stays at the station for 1 minute before starting for the next station. How many minutes does the train take to travel from first to the fifth station?

Answer: ....

PROBLEM 154

The human heart beats approximately 70 times per minute. How many beats approximately will it beat in an hour?

Answer: ....

Jenny, Kim, Sally and Lily were born on March 31st, May 18th, July 19th and March 19th. Kim and Sally were born in the same month. Jenny's and Sally's birthdays fall on the same dates in different months. Who was born on May 18th?

Answer: ....

PROBLEM 156

Mary leaves her house at 7:55 a.m. and reaches school at 8:30 a.m. Her friend Jenny arrives at school at 8:45 a.m. It takes her 15 minutes less than Mary to get there. When does Jenny leave her house?

Answer: ....

PROBLEM 157

Cindy is 45 years old now. Three years ago, her brother was 35 years old. What is their total age now?

Answer: ....

PROBLEM 158

Jimmy is 24 years old. His sister is a quarter of his age. How old is Jimmy's sister?

Answer: ....

PROBLEM 159

William is 7 times as old as his son this year.
His son was 4 years old last year. How old
will his son be when William is 60 years old?

Answer: ....

PROBLEM 160

Kelly is 15 years older than Joe. Bobby is twice as old as Kelly. If Joe is 18 years old, what is the total age of the 3 persons?

Answer: ....

PROBLEM 161

Lina is 36 years old now. She will be 4 times as old as her youngest sister in 8 years time. How old is her youngest sister now?

Answer: ....

PROBLEM 162

Seven years ago, Jane was 23 years old. She is 6 times as old as Peter is now? How old is Peter now?

Answer: ....

PROBLEM 163

Susan is 4 times as old as Caroline this year. Caroline will be 10 years old next year. How old was Susan last year?

Answer: …………………………

In the picture, the goat has a weight of 4 kg. What is the total weight of 4 rabbits and 8 cats?

Answer: ………………………

PROBLEM 165

What is the weight of one banana?

Answer: ....

PROBLEM 166

Joe weighs 9 kg more than Holly. Holly weighs 27 kg. How much does Joe weigh?

Answer: ....

PROBLEM 167

If an apple weighs 60 grams, how much will 5 apples weigh? How many apples do I need to make 540 grams?

Answer: …………………………

Darren used to weigh 42 kg. After Darren lost some weight, Joseph weighed 6 kg more than him. If Joseph's weight is 38 kg, how much did Darren lose?

Answer: ....

PROBLEM 169

Caroline bought 1 kg of rice. She used 480 grams and packed the rest equally into 4 bags. How much rice was there in each bag?

Answer: ....

PROBLEM 170

Edward and Henry weigh 93 kg together.
Edward and Paul weigh 87 kg together. If
Edward weighs 36 kg, how much do Henry
and Paul weigh?

Answer: ....

PROBLEM 171

I want to bake 12 cakes. If we know that 6 kg of flour is enough for 36 cakes, how much flour do I need?

Answer: ....

PROBLEM 172

Melvin is 4 times as heavy as Peter. If Peter is 26 kg;

a) How much does Melvin weigh?

b) How much heavier is Melvin than Peter?

Answer: ....

PROBLEM 173

Two bags of cotton weigh 396 grams altogether. After 33 grams of the cotton from Bag A was transferred to Bag B, both bags weigh the same. What was the weight of bag A at the start?

Answer: ....

PROBLEM 174

Two equal pieces of string are tied together. 10 cm from each piece gets used up in tying the knot. After tying, the combined length of the two strings is 80 cm. How long was each piece of string before they were tied up?

80 cm

Answer: ....

PROBLEM 175

There are ten trees in front of John's house. The distance between the first tree and second tree is 1 m, the second tree and third tree is 2 m, the third tree and fourth tree is 3 m, and so on. What is the distance between the first and the last tree?

Answer: ....

Claire bought a piece of cloth that was 350 centimeters long. She cut it into 5 equal pieces and used 3 such pieces. Find the length of cloth that she had left.

Answer: ....

PROBLEM 177

Anne has to travel a distance of 20 km. She travels 15 km on a train and 3.5 km on a bus. She walks the rest of the way. How far does she have to walk?

Answer: ....

PROBLEM 178

Dennis traveled from his home to the beach. He cycled for 3 km and walked 1 km less than the distance he cycled. He then took a bus for the rest of the journey. If the distance from his house to the beach is 8 km, what was the distance Dennis traveled by bus?

Answer: ....

PROBLEM 179

A ribbon is 18 meters long. A rope is 125 centimeters long. How much longer is the ribbon than the rope in meters and centimeters?

Answer: ………………………….

PROBLEM 180

Pete went swimming. Each length of the pool was 50 meters long. He swam 6 lengths. How many more lengths does he have to swim so that he has done 500 meters in the pool?

Answer: ....

PROBLEM 181

I have 9 meters of material. I need to cut lengths of 30 centimeters. How many complete lengths can I cut? How much will be left over?

Answer: ....

PROBLEM 182

Dad needed 7 m of wood to build some shelves. He already had 125 cm of wood. How much more did he need to buy?

Answer: ....

If a snail travels 3 cm in 5 minutes, how far will it travel in half an hour?

Answer: ....

PROBLEM 184

My hybrid car travels 30 km for every one liter of fuel I fill in it. One liter of fuel costs $2. How far can I drive for $12?

Answer: ....

PROBLEM 185

Containers A, B and C are half-filled with water. There is 890 ml of water in container A, 120 ml of water in container B and 345 ml of water in container C. If the 3 containers are completely filled, what is the total volume that the 3 containers can hold?

Answer: ....

PROBLEM 186

An empty container was completely filled with water poured from 3 similar jugs. If 5 such jugs contained 400 ml of water, how much water was there in the container?

Answer: ....

Jeff mixed 840 liters of cold water to 196 liters of lemon concentrate to make lemonade. He then poured out 395 liters of the lemonade to serve his guests. How many liters of lemonade did he have left?

Answer: ....

PROBLEM 188

When a bucket is full it holds exactly 5 liters of water. A jug holds 500 milliliters. How many full jugs of water will I need to fill an empty bucket?

Answer: ....

Observe the block shown below. The outside of the block has been made with red cubes but the inside part has been made with blue cubes. How many blue cubes have been put inside this block?

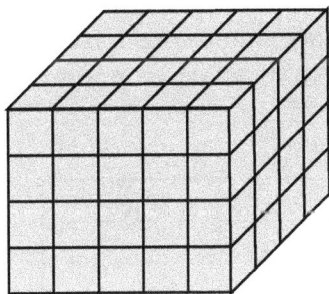

Answer: ....

A rectangle has an area 72 sq.cm.

a) Complete the table below to give the possible dimensions of the rectangle.

b) What are the dimensions of the rectangle with the largest perimeter?

c) What are the dimensions of the rectangle with the smallest perimeter?

Width	Length
1 cm	
2 cm	
3 cm	
4 cm	

Answer: ....

Lisa drew the shape shown below. She calculated the area of her shape.

Area = 48 sq.cm.

- Write the measurement of each side on the lines below.
- What is the perimeter of Lisa's figure.

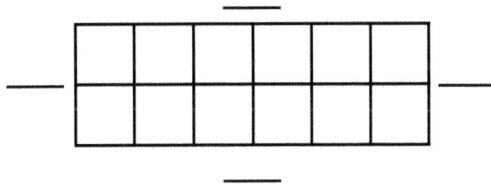

Answer: ....

PROBLEM 192

A jogger runs once around a square field.
Each side of the field measures 6500 cm.

a) How far does the jogger run altogether?
   Write your answer in meters.

b) What distance did the jogger run if he
   went four and a half times around the
   field? Write your answer in meters.

c) How many times would the jogger have
   to run around the field to run a distance
   of 2600 meters?

Answer: ....

The figure below shows a field. The length of the field is 3 times its breadth. It costs $8 per meter to put a fence around the field. How much will it cost to fence the entire field?

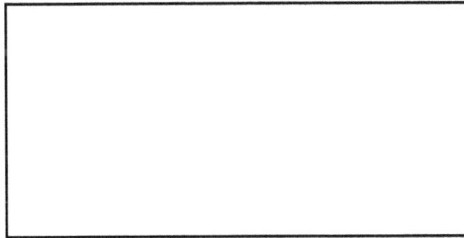

24 m

Answer: ....

What is the maximum number of squares of sides 2 cm that can be cut from a rectangular sheet of paper shown below?

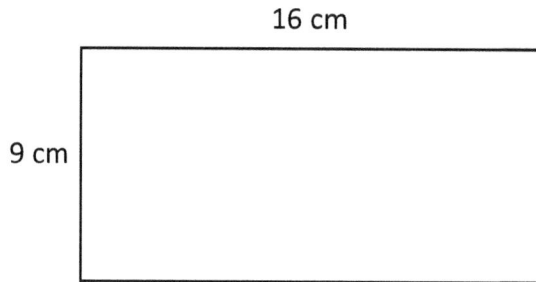

16 cm

9 cm

Answer: ....

The figure below is made up of 2 similar squares and 4 similar rectangles. Find the perimeter of the figure.

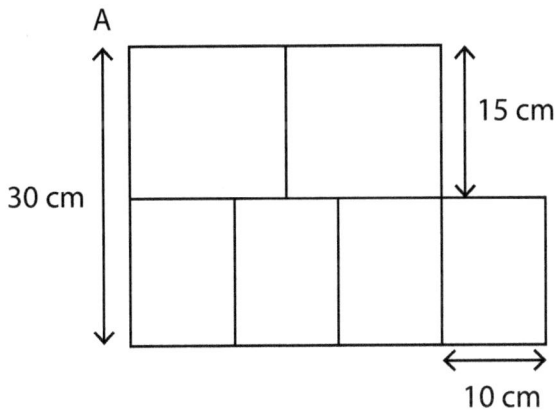

Answer: ....

The figure is made up of 2 similar rectangles and a square. Find the area of each rectangle.

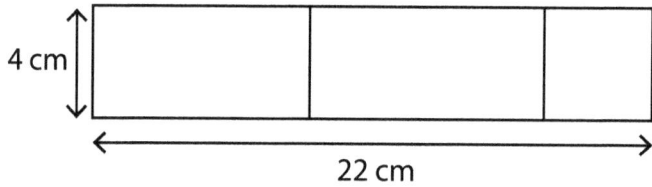

4 cm

22 cm

Answer: ....

A wire is cut and bent to form 5 similar rectangles. The length of each rectangle is twice its breadth. The length of a rectangle is 24 cm. What is the original length of the wire?

Answer: ....

Half of a piece of wire is bent to form a square and a rectangle shown below. What is the length of the wire?

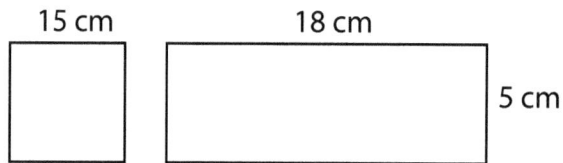

15 cm   18 cm

5 cm

Answer: ....

PROBLEM 199

A room measuring 12 m by 8 m is to be tiled. If it costs $5 per square meter to tile the room, what is the total cost of tiling the room?

Answer: ....

A rectangular plot of land measures 45 meters by 9 meters. A swimming pool with area of 250 meter square covers a portion of the land and a square grass patch with sides 8 meters covers another portion of the land. What is the area of the land that is not covered by the swimming pool and the grass patch?

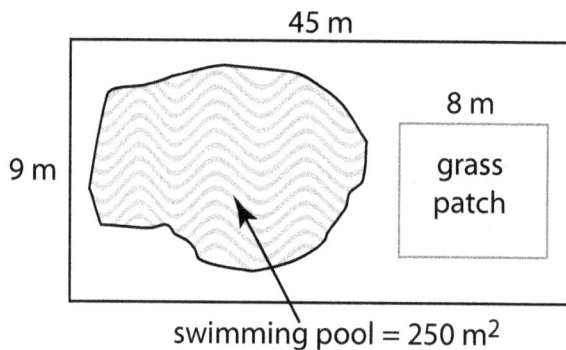

45 m

8 m

9 m

grass patch

swimming pool = 250 m^2

Answer: ....

SOLUTIONS

Solution to Question 1

I am a four digit number.
I am smaller than 6000 but bigger than 5000.
Therefore the digit at thousands place will be 5.
My hundreds digit is smaller than 8 but bigger than 6 = 7
8 > 7 > 6
My tens digit is an odd number; smaller than 5 but bigger than 1 = 3
The units digit is in the 3 times table and is bigger than 4 but smaller than 8.
= 3 x 2 = 6
I am 5736.

Solution to Question 2

I am a three digit number.
My units digit is 72 less than 81 = 81 -72 = 9
The tens digit is the same as 3 + 4 = 7
The hundreds digit is an odd number which is bigger than 1 but smaller than 4 = 3
I am 379.

Solution to Question 3

The number has five digits.
The ten thousands digit is the same as the hundreds digit, and it is an odd
number which can be divided by 3 to make 3 = 3 x 3 = 9
The thousands digit is one bigger than 8 = 9
The tens digit is smaller than 1 = 0
The units digit is 2 more than the tens digit = 0 + 2 = 2
The number is 99902.

Solution to Question 4

The number has five digits.
The ten thousands digit is 1 + 2 + 3 + 6 − 7 = 12 − 7 = 5
The thousands digit is the number of days in a week = 7
The hundreds digit is 4 more than the tens digit = 5 + 4 = 9
The tens digit is the number of fingers on your hand = 5
The units digit is half of 8 = 8/2 = 4
The number is 57954.

Solution to Question 5

Number of times digit nine appears in numbers from 1 to 100 = 20

9	91
19	92
29	93
39	94
49	95
59	96
69	97
79	98
89	99
90	

Note, In 99, 9 appears twice.

Solution to Question 6

The digits given are 3, 8, 6 and 9.
The smallest three digit number that you can make is 368. The largest two digit number you can make is 98.

The difference between the smallest three digit and the largest two digit number is 368 - 98 = 270

Solution to Question 7

The number between 160 and 190 is 185

Then number between 550 and 600 is 581

The sum of these numbers is 185 + 581 = 766

Solution to Question 8

a) The smallest number you can make using the three cards = 579
b) The number closest to 760 you can make using the three cards = 759

Solution to Question 9

a) About four million three hundred thousand children watch American Idol each month = 4,300,000
b) High School Musical sold two hundred and thirty thousand five hundred and sixty copies of their DVD last year = 230,560
c) Taylor Swift has sung two hundred and twenty two thousand, six hundred and five songs so far = 222,605

Solution to Question 10

The largest two-digit even number between 80 and 90 is 88.
The smallest two-digit even number between 20 and 30 is 22

The sum of 88 and 22 is 110

Solution to Question 11

A number divisible by 5 has to end in a 5.
The number should be largest three-digit number so it must start with a 9.
Since all digits should be different, the middle digit should be 8.
The number is 985.

Solution to Question 12

Digits that are divisible by 3 are 0, 3, 6 and 9
Since digits should be non-zero, that means we can not use 0 Also,
all digits are different, that means we have to use all the three digits
3, 6 and 9
We have to make the smallest number so the answer is:
369

Solution to Question 13

Lets work it out backwards.
If you take away one fourth of me, then add two, you get 17.
So if we remove 2 from 17, we will get = 17 - 2 = 15
15 is obtained after we take away one fourth of the number.
This means 15 is three fourths of the original number.
That means the number is (15/3) x 4 =20
I am 20.
Check:
5 + 5 + 5 + 5
1/4 is removed and 2 is added
= 5 + 5 + 5 + 2
= 17

Solution to Question 14

What number am I?
a) I am four less than 21 = 21 − 4 = 17
b) I am half of 26 = 26/2 = 13
c) I am a quarter of 4 = 4/4 = 1
d) I am a third of 9 added to 12 = 9/3 + 12 = 3 + 12 = 15
e) If you add 20 to me you get 50 = 50 − 20 = 30

Solution to Question 15

Find a pair of numbers with:

a) a sum of 11 and a product of 24.

$24 = 2 \times 12$
$= 3 \times 8$
$= 4 \times 6$

Out of these combinations only 3 + 8 = 11. Therefore the pair of numbers that gives a sum of 11 and product of 24 is 8 and 3

b) sum of 40 and a product of 400.

$400 = 20 \times 20$
$= 10 \times 40$
$= 8 \times 50$
$= 5 \times 80$
$= 4 \times 100$

Out of these combinations only 20 + 20 = 40. Therefore the pair of numbers that gives a sum of 40 and product of 400 is 20 and 20.

c) a sum of 15 and a product of 54

$54 = 18 \times 3$
$= 2 \times 27$
$= 6 \times 9$

Out of these combinations only 6 + 9 = 15. Therefore the pair of numbers that gives a sum of 15 and product of 54 is 6 and 9.

Solution to Question 16

In a street, houses are numbered with the consecutive odd numbers from 1 to 21 on one side.

1, 3, 5, 7, 9, 11, 13, 15, 17, 19, 21

Total number of houses that are numbered with odd numbers = 11

On the other side of the street, the houses are numbered with the consecutive even numbers from 2 to 16.

2, 4, 6, 8, 10, 12, 14, 16

Total number of houses that are numbered with even numbers = 8

Number of houses in the street altogether = 11 + 8 = 19

212

Solution to Question 17

Since there are 4 <u>consecutive odd numbers</u> that add up to 40, that means
they will be two more than and two less than 40/ 4 = 10
Two odd numbers lower than 10 and two higher than 10 are 7, 9, 11 and 13
Check: 7 + 9 + 11 + 13 = 40
The largest of these 4 odd numbers is 13.

Solution to Question 18

1 ball point pen costs as much as 4 pencils.
1 ball point pen = 4 pencils
One fountain pen costs as much as 2 ball point pens.
1 fountain pen = 2 ball point pens
= 2 x 4 pencils
= 8 pencils
Number of pencils that you can buy with the money for one fountain pen = 8

Solution to Question 19

 8 boys wore spectacles and 5 more girls than boys wore spectacles.
All the girls in the class wore spectacles
Number of girls who wore spectacles = 8 + 5 = 13
There were 15 more boys than girls in a class.
Number of boys in the class = 13 + 15 = 28
Number of children in the class = 28 + 13 = 41

Solution to Question 20

1 week = 7 days
2 weeks = 2 x 7 = 14 days

Tony eats a 1/4 bag of popcorns every day.
Therefore in 4 days, number of bags eaten = 1/4 x 4 = 1 bag

4 days ——— 1 bag
14 days ——— ? Bag

= 14/4
= 3.5 bags

Number of bags of popcorns Tony would need to have enough popcorn for two weeks = 3.5 bags i.e. three full bags and one half bag.

Solution to Question 21

The next palindrome number will be 212.
Sum = 202 + 212 = 414
The sum of 202 and the next palindrome number 212 is 414.

Solution to Question 22

Jenny is reading a story book.
The sum of the two facing pages that she is reading is 37. The number on the right side is one more than the number on the left side of the page.
Left side page number + right side page number = 37
Left side page number + (left side page number + 1) = 37
2 x left side page number = 37 - 1 = 36
Left side page number = 18 and Right side page number = 18 + 1 = 19
The two facing page numbers are 18 and 19.

Solution to Question 23

Jim placed 12 cans in a row.
He had fifty $1 coins.
He put one $1 coin in the first can, two $1 coin in the second can, three $1 coin in the third can.
From the above pattern, it is clear that the number of coins being put in every consecutive can is increasing by 1.
When 14 coins are left with Jim the number of coins used = 50 − 14 = 36
Count the number of cans to get a sum of consecutive numbers 36.
= 1 + 2 + 3 + 4 + 5 + 6 + 7 + 8
Number of cans used when Jim had 14 coins left with him = 8 cans

Solution to Question 24

Jason used some coins to make a triangle.
He placed one coin on each corner of the triangle.
There are 5 coins on each side of the triangle.
A triangle has 3 sides.
Number of coins Mark used = 5 + 4 + 3 = 12 coins

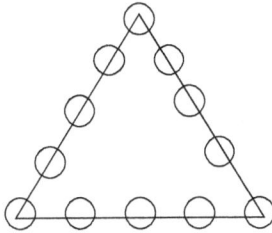

Solution to Question 25

Jenny puts 9 coins in a row on a table. Each coin is 10 cm apart.
The total number of spaces between the coins will be 8.
Therefore the ninth coin is 8 x 10 = 80 cm from the first one.

Solution to Question 26

The houses in a street are numbered sequentially from 1 to 24.
2, 12, 20, 21, 22, 23, 24
Number of times digit 2 occurs in the numbers of those houses = 8

Solution to Question 27

Mark has 61 matches.
Using some of them he made a triangle whose sides had 7 matches each.
A triangle has 3 sides
Total number of matches used for triangle = 7 + 7 + 7 = 21
Remaining matches = 61 – 21 = 40

With the remaining he made a rectangle whose one side had 6 matches.
A rectangle has 4 sides where 2 sides are of same length and the other 2 sides
are of equal length.
Number of matchsticks used for two sides = 6 + 6 = 12
Number of matchsticks used for the remaining two sides = 40 − 12 = 28
Therefore number of matchsticks used for 1 side = 28/2 = 14
The other side of the rectangle was 14 matchsticks long.

Solution to Question 28

A clock cost $24 and a watch cost $15 more than the clock.
Cost of the Watch = 24 + 15 = $39
Daniel bought 2 clocks and a watch.
Cost of 2 clocks and a watch = 2 x 24 + 39
= 48 + 39
= $87
Amount of money Daniel spent = $87

Solution to Question 29

There were 206 men at a table tennis match.
There were 49 fewer men than women.

a) Number of women at the match = 206 + 49 = 255
b) Number of people altogether = 255 + 206 = 461

Solution to Question 30

Jenny traveled 150 km by car.
50 km by bus.
3250 km by plane.
Total distance from her house = 150 + 50 + 3250 = 3450 km

Solution to Question 31

Joshua is 12 years old now. His father is 32 years older than him.

Father's age = 12 + 32 = 44 years
Joshua's age in 5 years time = 12 + 5 = 17 years
Fathers' age in 5 years would be = 44 + 5 = 49 years
Total ages of Joshua and his father in 5 years time = 49 + 17 = 66 years

Solution to Question 32

Jenny has 230 oranges.
Sarah has 145 oranges more than Jenny = 230 + 145 = 375
Jason has 310 oranges more than Sarah = 375 + 310 = 685
Number of oranges with the three children = 230 + 375 + 685 = 1290

Solution to Question 33

985 participants in a 5km cross country race are females.
The number of male participants is 1297 more than the number of female participants.
= 985 + 1297
=2282
Number of participants altogether = 985 + 2282 = 3267

Solution to Question 34

Regina had 1050 magazines.
Sophia had 672 fewer magazines than Regina.
Number of magazines Sophia has = 1050 – 672 = 378

Solution to Question 35

Sarah and Ben have 3540 stamps.
Sarah had 978 stamps.
Ben therefore had = 3540 – 978 = 2562 stamps
Difference = 2562 - 978 = 1584
Sarah had 1584 stamps fewer than Ben.

Solution to Question 36

There are 389 cookies in a box.
The box can hold 3160 cookies.
Number of cookies Shirley can put in the box = 3160 − 389 = 2771

Solution to Question 37

There are 200 circus members altogether and 60 are clowns.
There are 20 less acrobats than clowns = 60 − 20 = 40
Remaining = 200 − 60 − 40
= 200 − 100
= 100
There are same number of elephants and tigers in the circus = 50 + 50
Number of elephants = 50
Number of tigers = 50
Number of acrobats = 40

Solution to Question 38

Cherry saved $190 in a week.
She saved $48 on Monday and $24 on Tuesday.
Amount of money she saved for the rest of the week = 190 − 48 − 24 = $118

Solution to Question 39

A bus leaves at 16.20 p.m. with 24 people traveling in it and reaches Harbor
Centre at 17.52 p.m.
17 people get off the bus and 8 get on at Harbor Centre.
Total = 24 − 17 + 8 = 15
Number of people still on the bus at 17.54 p.m. = 15

Solution to Question 40

The difference between two numbers is 150.
The smaller number is 2356.

218

The bigger number = 2356 + 150 = 2506
Sum of the two numbers = 2506 + 2356 = 4862

Solution to Question 41

After Jeffrey gave 20 sweets to Adeline, she had 70 sweets and he had 10 sweets.
Number of sweets Jeffrey had at first = 10 + 20 = 30
Number of sweets Adeline had at first = 70 − 20 = 50

Solution to Question 42

In a car park, there are 198 buses.
There were 54 more buses than vans.
Vans = 198 − 54 = 144
There are 36 more motorcycles than vans.
Number of motor cycles = 144 + 36 = 180
Total number of vehicles altogether = 198 + 144 + 180 = 522

Solution to Question 43

$1 can be exchanged with two 50 cent coins.
For $100 = 2 x 100 = 200 coins of 50 cents can be exchanged.

Solution to Question 44

Richard has 349 paper clips.
Kevin had 7 times as many paper clips as Richard.
= 7 x 349 = 2443
Number of paper clips Kevin had = 2443

Solution to Question 45

Caroline baked 7 trays of cupcakes.
There were 16 butter cupcakes and 10 chocolate cupcakes on each tray.

Number of butter cupcakes in 7 trays = 16 x 7 = 112
Number of chocolate cupcakes in 7 trays = 10 x 7 = 70
Number of butter cupcakes Caroline baked more than the chocolate cupcakes
= 112 – 70 = 42

Solution to Question 46

Samuel sold 152 oranges on Monday and 62 oranges on Tuesday.
Total oranges sold = 152 + 62 = 214
He had twice as many oranges left as the total number of oranges sold.
Number of oranges left = 2 x 214 = 428

Solution to Question 47

There are 328 stamps in 1 album.
Number of stamps in 6 such albums = 328 x 6 = 1968

Solution to Question 48

There are 27 boys and 15 girls in a class.
Total number of students = 27 + 15 = 42
Each student has 8 exercise books.
Number of exercise books they have altogether = 42 x 8 = 336

Solution to Question 49

One bag can contain 168 sweets.
The total number of sweets in 7 bags = 7 x 168 = 1176

Solution to Question 50

There were 110 black chairs.
There were 4 times as many white chairs as black chairs in an office. Number. of white chairs = 4 x 110 = 440
The white chairs were arranged in rows of 8.
Number of rows of chairs = 440/8 = 55

Solution to Question 51

There were 220 marbles in a box.
80 of them were red and the rest were green and yellow.
Number of red marbles = 80
Remaining (green + yellow) = 220 − 80 = 140
There were twice as many red as green marbles
Green marbles = 80/2 = 40
Yellow marbles = 140 − 40 = 100

Solution to Question 52

There are 5 times as many mangoes in box A than in box B. Ratio is 5: 1
There are 60 mangoes in both the boxes.
Number of mangoes in box B = 60 x 1/(1+5) = 60/6 = 10
Number of mangoes in box A = 10 x 5 = 50

Solution to Question 53

A box of 10 eggs cost $8
Two boxes would cost = 8 x 2 = 16
If the eggs were bought individually, they cost $2 each.
Cost of 20 eggs individually = 20 x 2 = $40
Difference in the cost = 40 − 16 = $24
It would cost $24 more to buy 20 eggs individually than buying 2 boxes of eggs.

Solution to Question 54

Dennis had 310 apples and 230 oranges.
Total fruits = 310 + 230 = 540
He sold twice as many apples as oranges and had 240 fruit left.
Fruits sold = 540 − 240 = 300
Since number of apples sold is twice the number of oranges sold, ratio = 2:1
Number of oranges sold = 300 x 1 / (1+2) = 300/3 = 100
Number of apples sold = 200 (twice of 100)

Solution to Question 55

A buffet lunch cost $23 for an adult and $7 less for a child
Cost of lunch for a child = 23 − 7 = $16
The total amount of money that a family of 2 adults and 3 children had to pay for the buffet lunch
= 2 x 23 + 3 x 16
= 46 + 48
= $94

Solution to Question 56

There were 110 green stickers and twice as many blue stickers as green stickers. Number of blue stickers = 110 x 2 = 220
There were 85 red stickers.
Number of stickers altogether = 110 + 220 + 85 = 415

Solution to Question 57

Marcus had 120 red balloons and 210 green balloons.
Total balloons = 120 + 210 = 330
Jacquelyn had twice as many balloons as him and Dennis had 3 times as many balloons as him.
Number of balloons Jacquelyn had = 2 x 330 = 660
Number of balloons Dennis had = 3 x 330 = 990
Number of balloons altogether = 330 +660 +990 = 1980

Solution to Question 58

There were 120 apples and oranges in a box.
Since the number of oranges was twice the number of apples in a box;
Ratio of number of oranges to the number of apples = 2 : 1

Number of apples = 120 x 1/ (1 + 2) = 120 / 3 = 40

Number of oranges = 40 x 2 = 80

Solution to Question 59

Gavin arranged 25 rows of 7 chairs
Total Number of chairs = 25 x 7 = 175
137 chairs were occupied.
Number of chairs not occupied = 175 − 137 = 38

Solution to Question 60

There were a total of 280 students, comprising of Americans, Chinese and Indians.
40 Americans left the basket ball tournament, and the remaining students were Chinese and Indians.
Number of students remaining = 280 − 40 = 240
As the number of Chinese students was 5 times more than the number of Indians, ratio = 5 : 1
Number of Indian students = 240 x 1 (1 + 5) = 240 / 6 = 40
Number of Chinese students = 40 x 5 = 200
Difference = 200 − 40 = 160
There were 160 more Chinese students than Indian students.

Solution to Question 61

In 1 box, there were 175 cookies.
Margaret had 6 such boxes of cookies.
Total cookies = 6 x 175 = 1050
She sold 368 cookies to Kelvin.
Number of cookies she had left = 1050 − 368 = 682

Solution to Question 62

Jason bought 4 pairs of shoes at $58 each.
Cost of shoes = 58 x 4 = $232
He bought 2 pairs of socks at $5 each.
Cost of socks = 5 x 2 = $10
He then had $85 left.
Money he had at first = 232 +10 + 85 = $327

Solution to Question 63

Mary has 95 books.
David has thrice as many books as Mary = 95 x 3 = 285
Rose has the same number of books as David = 285
a) Number of books David has = 285
b) Number of books all the three children have altogether = 95 + 285 + 285 = 665

Solution to Question 64

After giving 128 marbles to each of the 7 boys, Mrs Smith had 34 marbles left
a) Number of marbles the 7 boys received altogether = 128 x 7 = 896
b) Number of marbles Mrs Chan had at first = 896 + 34 = 930

Solution to Question 65

Ahmad arranged 158 stamps into each of the 5 albums.
Total number of stamps = 158 x 5 = 790
 Ahmad had 97 stamps left.
= 790 + 97 = 887
Number of stamps Ahmad had at first = 887

Solution to Question 66

Kelvin spent $360.
Jane spent half as much as him = 360/2 = $180
Amount of money they spent altogether = 360 + 180 = $540

Solution to Question 67

104 children were going to the zoo in minibuses.
Each minibus could seat 8 children.
The least number of minibuses needed to take all the children to the zoo = 104/8
= 13

Solution to Question 68

Mrs. Thomas baked 104 apple pies and 216 lemon pies.
She packed the two different types of pies separately into boxes of 4 each.
Number of apple pie boxes = 104/4 = 26
Number of lemon pie boxes = 216/4 = 54
Difference = 54 − 26 = 28
There were 28 more boxes of lemon pies than apple pies.

Solution to Question 69

Claire has 840 sweets.
She has 4 times as many sweets as Caroline.
Number of sweets Caroline has = 840/4 = 210
Number of sweets they have altogether = 210 + 840 = 1050

Solution to Question 70

There were 10 balloons in a packet.
Caroline bought 8 such packets.
Total balloons in 8 packets = 8 x 10 = 80
She blew them and tied them into bunches of 5.
Number of bunches of balloons = 80/5 = 16

Solution to Question 71

2 similar books cost $186.
Cost of each book = 186/2 = $93
Doris bought 3 books and had $28 left. Amount
of money she had at first   = 3 x 93 + 28

= 279 + 28 = $307

Solution to Question 72

Doris bought 8 packets of tomatoes and gave them to 25 students.
When she gave each of them 4 tomatoes, she had 4 tomatoes left.
Total tomatoes given to students = 25 x 4 = 100

Tomatoes left with her = 4
Total tomatoes = 100 + 4 = 104
Number. of tomatoes in each packet = 104/8 = 13 tomatoes

Solution to Question 73

A shopkeeper sold 230 potatoes on Monday and twice as many potatoes on Tuesday.
Number. of potatoes sold on Tuesday = 2 x 230 = 460
He had 150 potatoes left.
Total number of potatoes = 460 + 230 + 150 = 840
He wanted to sell equal number of potatoes within 2 days.
Number of potatoes he should sell on each day = 840/2 = 420

Solution to Question 74

9 packets of sweets cost $72 and 3 packets of biscuits cost $12.
1 packet of sweets = 72/9 = $8
1 packet of biscuits = 12/3 = $4
Amie wanted to buy 4 packets of sweets and 2 packets of biscuits.
= 4 x 8 + 2 x 4
= 32 + 8
= $40
Total amount of money that Amie had to pay for the items = $40

Solution to Question 75

Florence bought 18 packets of apples weighing 5 kg each.
Total weight of apples = 18 x 5 = 90kg
She then packed them into plastic bags of 2 kg each.
Number. of plastic bags she used = 90/2 = 45

Solution to Question 76

Henry had 542 carrots.
He sold 138 carrots

Number of carrots remaining = 542 − 138 = 404
He packed the remaining carrots equally into 4 boxes.
Number of carrots in each box = 404/4 = 101

Solution to Question 77

8 people paid $90 each for a dinner.
Amount of money paid by 8 people = 90 x 8 = $720
The organizers were still short of $24.
Actual amount that 8 people should have paid = 720 + 24 = $744
The actual amount of money that each of them should pay = 744/8 = $93

Solution to Question 78

There were 250 fiction books and 180 non-fiction books in a library.
There were 3 times as many non-fiction books as scientific books.
Number of science books = 180/3 = 60
Total number of books altogether = 250 + 180 + 60 = 490

Solution to Question 79

Tracy bought 2 similar watches and had $150 left.
She spent 3 times the amount of money left with her.
Money spent = 150 x 3 = $450
Cost of each watch = 450/2 = $225

Solution to Question 80

Grade 3 is having a pizza party.
The teacher ordered 6 cheese pizzas, 9 chicken pizzas, and 6 vegetarian pizzas.
Total number of pizzas = 6 + 9 + 6 = 21
Each pizza is divided into 6 pieces
Total number of pieces = 21 x 6 = 126
Each student gets 3 pieces
Number of students = 126/3 = 42

Solution to Question 81

There were 24 potatoes in a box.
Thomas bought 2 boxes
Number of potatoes in 2 boxes = 24 x 2 = 48
He used 12 potatoes.
Remaining potatoes = 48 – 12 = 36
He then repacked the remaining potatoes into packets of 6.
Number of packets of potatoes he had = 36/6 = 6

Solution to Question 82

An adult's ticket to the zoo safari cost $18.50.
A child's ticket cost $7.50.
Florence took her three children to the zoo.
Cost of tickets = 18.5 + 3 x 7.5 = 18.5 + 22.5 = 41
After paying for all the tickets, she had $56 left.
Amount of money she had at first = 56 + 41 = $97

Solution to Question 83

There were 8 charity tickets in a booklet.
Linda sold 18 booklets and Jennifer sold 4 times as many booklets as her.
Number of tickets Linda sold = 18 x 8 = 144
Number of tickets Jennifer sold = 4 x 18 x 8 = 576
Michael sold half the number of booklets that Linda sold.
Number of tickets Michael sold = 18/2 x 8 = 72
Number of charity tickets they sold altogether = 144 + 576 + 72 = 792

Solution to Question 84

Caroline had 6 times as many sweets as Melvin. They had a total of 28 sweets.
Melvin had 28 x 1 / (1 + 6) = 4 sweets
This means Caroline had 6 x 4 = 24 sweets
If they divide sweets equally between them each one should have half of 28 sweets
= 28/2 = 14 sweets

Number of sweets Caroline should give to Melvin so that they have the same number of sweets = 24 − 14 = 10

Solution to Question 85

Box A contained 250 marbles and box B contained half as many marbles as box A.
Number of marbles in box B = 250/2 = 125
Box C contained 3 times as many marbles as box B
Number of marbles in box C = 3 x 125 = 375
Total number of marbles in the 3 boxes = 250 + 125 + 375 = 750

Solution to Question 86

A bottle of detergent cost $9 before the sale.
During the sale, the price of the bottle of detergent was $6.
Daniel bought 8 bottles of detergent before the sale.
Total cost of the detergent before the sale = 9 x 8 = $72
Number of bottles Daniel could buy with $72 during the sale = 72/6 = 12
Difference = 12 - 8 = 4
Daniel could buy 4 more of bottles of detergent during the sale.

Solution to Question 87

Jessica packed 23 boxes of 6 cupcakes each.
Number of cupcakes in 6 boxes = 23 x 6 = 138
She then had 3 cupcakes left.
Total number of cupcakes = 138 + 3 = 141
Number of Boxes Jessica would need if she packed cupcakes in boxes of 3 = 141/3 = 47
Difference in number of boxes = 47 − 23 = 24
Jessica would need 24 more boxes if she packed all the cupcakes into boxes of 3 instead.

Solution to Question 88

Betty has 149 buttons.
She needs to put the buttons equally into 6 drawers.
= 149/6 = 24 remainder 5

The buttons can be put into 6 drawers with 24 buttons in each drawer.
5 buttons will be left behind.

Solution to Question 89

It takes 20 minutes to cut a log into 5 pieces.
Time taken to cut 1 piece = 20/5 = 4 minutes
Time taken to cut 10 pieces = 10 x 4 = 40 minutes

Solution to Question 90

Peter wants to make the biggest rectangular yard with 24 meters of fence.
$2(L + B) = 24$
$L + B = 12$
Various combinations of length and breadth of the rectangular yard possible are as follows:
$12 = 8 + 4 = (8 \times 4 = 32)$
$= 6 + 6 = (6 \times 6 = 36)$
$= 10 + 2 = (10 \times 2 = 20)$
$= 9 + 3 = (9 \times 3 = 27)$
$= 7 + 4 = (7 \times 4 = 28)$
$= 5 + 7 = (5 \times 7 = 35)$
$= 11 + 1 = (11 \times 1 = 11)$

Since it's a rectangular yard the area cannot be 36 sq.m. The rectangular yard will have the maximum area when the length and the breadth are 7 meters and 5 meters.

Solution to Question 91

There are 132 students going for the movie.
Since each row has 11 seats, the number of rows that will be used by the students = 132 /11 = 12
The students will use 12 rows full of seats.

Solution to Question 92

There are 9 rows of 24 chairs in the school hall.
Total number of chairs = 9 x 24 = 216
Mr. Tim rearranged them into rows of 6.
Number of rows after Mr Tim rearranged the chairs = 216/6 = 36 rows

Solution to Question 93

A shopkeeper sold 234 potatoes on Monday and twice as many potatoes on Tuesday.
Number of potatoes sold on Tuesday = 2 x 230 = 460
He had 150 potatoes left.
Total number of potatoes = 468 + 234 + 150 = 852
He wanted to sell equal number of potatoes within 2 days.
Number of potatoes he should sell on each day = 852/2 = 426

Solution to Question 94

Jerry had 4 boxes of crayons and each box had 24 crayons.
Total number of crayons he had = 4 x 24 = 96 crayons.
Total number of crayons lost = 8 + 16 = 24
Remaining crayons with Jerry = 96 - 24 = 72
He puts these 72 crayons into 2 packets.
Each packet will have 72 / 2 = 36 crayons

Solution to Question 95

Total number of flowers that Alice picked up = 66 x 3 = 198
Alice made bunches of 12 flowers. Counting by 12's Alice will use:
12, 24, 36, 48, 60, 72, 84, 96, 108, 120, 132, 144, 156, 168, 180,192 flowers.
After that Alice will be left with 198 - 192 = 6 flowers that she will not be able to use in a bunch.
Alice will throw away 6 flowers.

Solution to Question 96

There were 28 blue plates and 29 green plates.
There were thrice the number of bowls as green plates.
Number of bowls = 29 x 3 = 87
Total number of plates and bowls = 28 + 29 + 87 = 144

Solution to Question 97

A third of the class has pets.
There are 45 pupils in the class.
Number of pupil who have pets = 45/3 = 15

Solution to Question 98

One fifth of the thirty five pupils in the drama club are girls.
Number of girls in the drama club = 35/5 = 7

Solution to Question 99

A cake is cut into fifteen equal pieces.
Sam eats five pieces at the break time.
Remaining number of cake pieces = 15 − 5 = 10
Fraction of the cake left = 10/15 = 2/3

Solution to Question 100

Two third students out of a class of 30 understand the math problem.
Fraction of the class that doesn't understand the math problem = 1 − 2/3
= (3 − 2)/3
= 1/3

Solution to Question 101

One quarter of the school walks to school.
Another quarter comes to school cycling.
Total = 1/4 + 1/4 = 1/2
Fraction of the school that does neither = 1 – 1/2
= 1/2

Solution to Question 102

The box had 12 chocolates.
One quarter of a box of chocolates has been eaten = 12/4 = 3
Remaining number of chocolates = 12 – 3 = 9
Fraction of chocolates left = 9/12 = 3/4

Solution to Question 103

Grade 3 has 40 students.
1/4 of the children have i-pads = 40/4 = 10
One half have an i-pod Touch = 40/2 = 20
a) Number of children who have i-pads = 10
b) Number of children who have an i-pod Touch = 20
c) Number of children in the class who have neither = 40 – 10 – 20
= 40 – 30
= 10

Solution to Question 104

Size of each piece	No. of pieces	Is this whole number of cms?	Fraction of whole
4 cm	40/4 = 10	Yes	1/10
8 cm	40/8 = 5	Yes	1/5
10 cm	40/10 = 4	Yes	1/4
20 cm	40/20 = 2	Yes	1/2

Solution to Question 105

Charles wants to put 15ml of fruit punch equally into glasses.
To get an exact whole number of ml in each glass is possible if the number of filled glasses is a factor of 15 i.e. 15 can be completely divided by the number of glasses.
Let's first write down the numbers by which 15 can be divided completely.
They are 1, 3, 5, 15

No. of glasses	Amount in each glass	Fraction of the whole
1	15/1 = 15 ml	1
3	15/3 = 5 ml	1/3
5	15/5 = 3 ml	1/5
15	15/15 = 1 ml	1/15

Solution to Question 106

A pizza has been divided into 8 equal slices.
Ben eats a quarter of the pizza.
This means he has eaten 1/4 of the pizza.
Tim eats another quarter = 1/4
Together they have eaten = 1/4 + 1/4 = 1/2 of the pizza Remaining fraction of the pizza= 1- 1/2 = 1/2
Number of slices left = 8/2 = 4

234

Solution to Question **107**

Moshi monster cards are selling for a discounted price; they normally cost $4.50 per packet but have been reduced to half price.
Cost of each packet now = 4.5/2 = $2.25

Solution to Question **108**

Peter had $80, but he spent $60.
Fraction of the money he spent = 60/80 = 3/4

Solution to Question **109**

I spent $69 at a book sale, and also spent two third of that amount to buy a present for my friend.
Amount I spent to buy the present = 2/3 x 69 = $46

Solution to Question **110**

In Grade 3, there are 30 girls and 20 boys.
Total number of students = 30 + 20 = 50
Fraction of girls in the class = 30/50 = 3/5

Solution to Question **111**

I ate 1/2 of a cake.
Remaining cake = 1 − 1/2 = 1/2
You eat 2/4 of a cake = 1/2
Cake left = 1/2 − 1/2 = 0

Solution to Question **112**

The recipe of baking one cake uses 1/2 cup sugar, 1/2 cup brown sugar, 1/4 cup flour, and 4 eggs.
To bake 2 cakes the quantity of each ingredient needed is:

Sugar = 2 x 1/2 cup = 1 cup
Brown sugar = 2 x 1/2 cup = 1 cup
Flour = 2 x 1/4 cup = 1/2 cup
Eggs = 2 x 4 = 8

Solution to Question 113

There are 8 pigs, 4 cows, and 6 roosters at the farm.
Total number of animals = 8 + 4 + 6 = 18
The fraction of each animal at the farm is:
Pigs = 8/18 = 4/9
Cows = 4/18 = 2/9
Roosters = 6/18 = 3/9 = 1/3

Solution to Question 114

There are 120 children performing in the school annual concert.
Half of the students are given 4 invitations for their family to come to the concert.
Number of students that were given 4 invitations = 120/2 = 60 students
Number of invitations sent = 60 x 4 = 240 invitations
The other half is given 2 invitations each.
Number of remaining invitations = 60 x 2 = 120 invitations
Number of people invited in all to see the concert = 240 + 120 = 360 invitations

Solution to Question 115

Tickets for Lady Gaga's concert are selling for $320.
They are on sale today for half price.
Now the tickets cost = 320/2 = $160

Solution to Question 116

William earned $3500 a month.
He spent $1420 on his children and $1200 on transport and food.
He then saved the rest of his money
Money saved every month = $3500 − $1420 − $1200 = $880
The amount he saved in 3 months = $880 x 3 = $2640

Solution to Question 117

A shirt cost $45 and a pair of jeans cost $52.
Total cost of both the items = 45 + 52 = $97
He was short of $13.
Amount of money with Samuel = 97 − 13 = $84

Solution to Question 118

An orange cost 1.50 dollars and a guava costs 50 cents more than
the orange. The guava costs = $1.50 + $0.50 = $2
Jessica bought an orange and a guava.
Total money spent = 1.50 + 2.00 = $3.50
She had $15 left.
Total amount of money Jessica had at first = $15 + $3.50
= $18.50

Solution to Question 119

Henry had $35.
He bought 5 boxes of pizza.
He had $10 left.
Cost of 5 boxes of Pizzas = 35-10 = 25
Cost of one box = 25/5 = 5
Cost of 8 boxes of pizza = 8 x 5 = $40

Solution to Question 120

Gavin earns $4350 every month.
He spends $350 on transport.
$120 more on food than on transport.
Amount of money spent on food = 350 + 120 = $470
He then gives $600 to his wife. He saves the rest.
Amount of money saved = 4350 − 350 − 470 − 600 = $2930

Solution to Question 121

7 books cost $105.
Cost of each book = 105/7 = $15
Victor bought 9 such books and had $14 left.
Amount of money he had at first = 9 x 15 + 14 = 135 + 14 = $149

Solution to Question 122

Richard spent $70.50 on a camera and $55.50 less on a radio.
Money spent on the Radio = $70.50 – $55.50 = $15
Total amount of money Richard spent = $70.50 + $15.00 = $85.50

Solution to Question 123

A toaster cost $87.
An oven cost $150 more than the toaster.
Cost of the oven = 150 + 87 = $237
A juicer cost $28 less than the oven.
Cost of the juicer = 237 – 28 = $209
Elizabeth bought the 3 electrical items and paid with a $1000 Note.
Amount of change she will get back = 1000 – 87 – 237 – 209 = $467

Solution to Question 124

Florence packed 100 apples equally into boxes of 5.
= 100/5 = 20 boxes
She sold each box for $3.
Amount of money Florence earned = 20 x 3 = $60
Oliver packed 100 oranges equally into boxes of 4.
= 100/4 = 25 boxes
He sold each box for $2.
Amount of money Oliver earned = 25 x 2 = $50
Difference between money earned by Florence and Oliver = 60 – 50 = $10
Florence earned $10 more than Oliver.

Solution to Question 125

Samuel saved $564 each month.
His total savings for 6 months = 6 x 564 = $3384
After 6 months, he spent some of his savings and had $1205 savings left.
Amount of money Samuel spent = 3384 − 1205 = $2179

Solution to Question 126

An electronic dictionary cost $78.
A cake mixer cost $22 more than the electronic dictionary.
Cost of the cake mixer = 78 + 22 = $100
A toaster costs half as much as the cake mixer = 100/2 = $50
Total cost of the 3 electrical items = 78 + 100 + 50 = $228

Solution to Question 127

Maggie had $290 at first. She gave $50 to Jeffrey.
Amount of money left with Maggie = 290 − 50 = $240
She still had thrice as much money as Jeffrey.
Amount of money with Jeffery = 240/3 = $80
Amount of money with Jeffery at first = 80 − 50 = $30

Solution to Question 128

Kelvin took his wife and 3 children to a restaurant for dinner. He
paid a total of $30.40.
Cost of each adult $8.00.
Cost of 2 adults = 2 x 8.00 = 16.00
Amount of money paid for 3 children = 30.40 − 16.00 = $14.40
Amount of money paid for 1 child = 14.40/3 = $4.80

Solution to Question 129

William spent $213.50 on Saturday.
He spent $89 more on Sunday than on Saturday.

Amount of money spent on Sunday = $213.50 + $89.00 = $302.50
Amount of money William spent on both the days = $213.5 + $302.5 =
$516.00

Solution to Question 130

Larry bought 9 English books and 3 Chinese books for a total of $330.
Each Chinese book cost $35.00
Cost of 3 Chinese books = 3 x $35.00 = $105.00
Cost of 9 English book = $330 – $105 = $225
Cost of 1 English book = $225/9 = $25
Cost of each English book = $25

Solution to Question 131

a) Susan bought 6 rings at $48 each and gave the cashier three hundred-dollar. Cost of 6 rings = $48 x 6 = $288
b) Amount of change Susan received from the cashier = $300 – $288 = $12

Solution to Question 132

Melvin bought 4 clocks and a T-shirt for $320.
Each clock cost $73.
Cost of 4 clocks = 4 x 73 = $292
292 + T-shirt = $320

a) Cost of the T-shirt = 320 – 292 = $28
b) Difference between the cost of the clock and the T-shirt = 73 – 28 = $45
   The clock costs $45 more than the T-shirt.

Solution to Question 133

Lester bought 8 chairs and a table for $1085.
 The table cost $389.
Cost of 8 chairs = $1085 – $389 = $696
a) Cost of 1 chair = $696/8 = $87
b) Amount of money Lester paid for one chair and a table put together =
   $87 + $389 = $476

Solution to Question 134

The total cost of a pencil case and a pencil is $2.40
The pencil case cost $1.00 more than the pencil. This means,
Cost of pencil + cost of pencil + $1.00 = $2.40
Cost of pencil + Cost of pencil = $1.40
So the cost of two pencils = $1.40 and cost of one pencil = $0.70
And the cost of the pencil case = $0.70 + $1.00 = $1.70

Solution to Question 135

$1 = 100 cents
$1 = 25 cents x 4. This means for $1, we get 4 twenty-five cent coins.
For $10, we will get 4 x 10 = 40 twenty-five cent coins.
Number of twenty-five cent coins that can be exchanged with $10 = 40

Solution to Question 136

Betty paid $16 for each of the 4 books she bought and had $10 left.
Cost of 4 books = 16 x 4 = $64
Amount of money left with Betty after buying 4 books = $10
Amount of money Betty had at first = 64 + 10 = $74

Solution to Question 137

Joe and Julia spent $13.75 each.
Amount of money they spent altogether = 13.75 x 2 = $27.5
Joe, Julia and Bobby spent $60.00 altogether.
Amount of money spent by Bobby = 60 – 27.5 = $32.5

Solution to Question 138

Mr. Tan bought 9 books at $8.00 each.
Total money spent = $8 x 9 = $72

He bought a ring for $68.50
Total amount of money he spent = 72 + 68.50 = $140.5

Solution to Question 139

A lamp costs $26.
An oven costs $89 more than the lamp = $26 + $89 = $115
Cost of the oven = $115

Solution to Question 140

Laura has $230.
She spent $120 on cosmetics.
She now had $60 left.
Amount of money spent on jewellery = 230 − 120 − 60 = $50

Solution to Question 141

Henry saves $285 a day.
Amount of money he will save in 9 days = 285 x 9 = $2565

Solution to Question 142

Cost of 500 grams sugar = 20 cents
Cost of 1000 grams or 1 kilogram sugar = 2 x 20 = 40 cents
Cost of 4500 grams or 4.5 kilograms of sugar = 40 + 40 + 40 + 40 + 20 = 180 cents
= $1.8

Solution to Question 143

Betty spent $269 on Monday and $195 more on Tuesday than on Monday.
Money spent on Tuesday = $269 + $1 95 = $464
Betty saved the rest of her money. She saved $1965.
Amount of money she spent altogether = $269 + $464 = $733
Amount of money Betty had at first = $733 + $1965 = $2698

Solution to Question 144

Total cost of the Math and the English book = $27
The Math book costs $3 more than the English book.
Cost of the Math book = Cost of the English book + $3
This means,
Cost of English book + $3 + Cost of English book = $27
Cost of the English book = $24/2 = $12
Cost of the Math book = $12 + $3 = $15

Solution to Question 145

Kevin saves $45.80 a day.
1 week = 7 days
Amount of money he can save in a week = $45.80 x 7 = $ 320.60

Solution to Question 146

A motorcycle costs $1078.
A lorry costs 9 times as much as a motorcycle.
Cost of the lorry = 1078 x 9 = $9702

Solution to Question 147

A pen costs $6.80.
A watch costs 4 times as much as a pen = 4 x $6.80 = $27.20
Cost of the watch = $27.20
Total cost of the watch and the pen = $27.20 + $6.80 = $34.00

Solution to Question 148

6 boys paid $174 for a present and had $36 left.
Amount of money each boy paid for the present = 174/6 = $29
Number of pens they could buy with the remaining amount of money if each
pen cost $4 = 36/4 = 9 pens

Solution to Question 149

Television program will begin at 9:00 p.m.
It takes Jake half an hour to do his chores.
He takes half an hour to finish his homework
Total time = 0.5 + 0.5 = 1 hr
Time at which Jake should begin to finish his homework and chores so that he can watch the television program = 9:00 − 1 hour = 8: 00 p.m.

Solution to Question 150

The solar eclipse began at 2:13 in the afternoon and ended at 3:51 p.m.
Duration of the solar eclipse = 3:51 p.m − 2:13 p.m. = 1 hour 38 minutes

Solution to Question 151

a) Tina's alarm rings at 6 o'clock. Four and a half hours later she leaves the house. Time at which Tina left = 6:00 + 4:30 = 10:30 a.m.
b) Tina and Ricky leave their home at 9 o'clock. They arrive at the park at 1:30 p.m. Time taken to walk to the park = 1:30 − 9:00 = 4 hours 30 minutes

Solution to Question 152

a) Time for which Ben and Mark had to wait until the shop opened = 11:30 - 9:15 = 2hours 15 minutes

b) Walk to the library took 20 minutes.
They read books for 30 minutes.
The time then was = 11:45 + 0:20 + 0:30 = 12:35 p.m.

Solution to Question 153

A train takes 4 minutes to reach the next station.
It stays at the station for 1 minute before starting for the next station.
Total time = 5 minutes

First station to second station = 5 min
Second station to third station = 5 min
Third station to fourth station = 5 min
Fourth station to fifth station = 4 min
Number of minutes that the train takes to travel from first to the fifth
station = 5 + 5 + 5 + 4
= 19 min

Solution to Question 154

The human heart beats approximately 70 times per minute.
1 hr = 60 minutes

1 min ——— 70 beats
60 min ——— ?
= 70 x 60
= 4200 beats
Number of beats that the human heart will make in an hour approximately =
4200

Solution to Question 155

Jenny, Kim, Sally and Lily were born on March 31st, May 18th, July 19th and March 19th.
Kim and Sally were born in the same month = March
Jenny's and Sally's birthdays fall on the same dates in different months = 19th
Lily was born on May 18th.

Solution to Question 156

Mary leaves her house at 7:55 a.m. and reaches school at 8:30 a.m.
Time taken by Mary = 8:30 − 7:55 = 35 minutes
Her friend Jenny arrives at school at 8:45 a.m. It takes her 15 minutes less than
Mary to get there.
Time taken by Jenny = 35 − 15 = 20 min
Jenny leaves her house at 8:45 − 0:20 = 8:25 a.m.

Solution to Question 157

Cindy is 45 years old now.
Three years ago, her brother was 35 years old.
Present age of her brother = 35 + 3 = 38 years
Their total age now = 45 + 38 = 83 years

Solution to Question 158

Jimmy is 24 years old.
His sister is a quarter of his age.
Jimmy's sister's age = 24/4 = 6 years

Solution to Question 159

William is 7 times as old as his son this year. His son was 4 years old last year.
Age of William's son this year = 4 + 1 = 5 years
Age of William this year = 7 x 5 = 35 years
Difference between William's age and his son's age = 35 -5 = 30 years
When William is 60 years his son would be = 60 – 30 = 30 years

Solution to Question 160

Joe is 18 years old.
Kelly is 15 years older than Joe = 18 + 15 = 33 years
Bobby is twice as old as Kelly = 33 + 33 = 66 years
Total age of the 3 people = 18 + 33 + 66 = 117 years

Solution to Question 161

Lina is 36 years old now.
She will be 4 times as old as her youngest sister in 8 years time.
After 8 years, Lina's age = 36 + 8 = 44 years
Her younger sister's age after 8 years = 44/4 = 11 years
Lina's youngest sister's age now = 11 – 8 = 3 years

246

Solution to Question **162**

Seven years ago, Jane was 23 years old.
Jane's present age = 23 + 7 = 30 years
She is 6 times as old as Peter is now.
Peter's age now = 30/6 = 5 years

Solution to Question **163**

Caroline will be 10 years old next year.
Caroline's present age = 9 years
Susan is 4 times as old as Caroline this year = 4 x 9 = 36 years
Susan's age last year = 35 years

Solution to Question **164**

In the picture, the goat has a weight of 4 kg.
1 goat = 2 rabbits
2 rabbits = 4 kg
1 rabbit = 4/2 = 2 kg
2 rabbits = 4 cats
But 2 rabbits = 4 kg
Therefore 4 cats = 4 kg
1 cat = 4/4 = 1 kg
Total weight of 4 rabbits = 4 x 2 = 8 kg
Total weight of 8 cats = 8 x 1 = 8 kg
Total weight of 4 rabbits and 8 cats = 8kg + 8kg = 16kg

Solution to Question **165**

2 apples = 90 g
1 apple = 90/2 = 45 g

3 pears = 4 apples
3 pears = 4 x 45 = 180 g
1 pear = 180/3 = 60 g

Weight of 1 banana = weight of 2 pears = 2 x 60 = 120 g

Solution to Question 166

Joe weighs 9 kg more than Holly.
Holly weighs 27 kg.
Joe weighs = 27 + 9 = 36 kg

Solution to Question 167

An apple weighs 60 g.
5 apples will weigh = 5 x 60 = 300 g
Number of apples I need to make 540 g = 540/60 = 9

Solution to Question 168

Darren was 42 kg. Joseph was 38 kg.
After Darren lost some weight, Joseph was 6 kg heavier than him.
Darren's weight now = 38kg - 6 kg = 32 kg
Darren's total weight loss = 42kg - 32 kg = 10 kg

Solution to Question 169

Caroline bought 1 kg of rice = 1000 grams of rice
She used 480 grams.
Remaining = 1000 – 480 = 520 grams
She packed the rest equally into 4 bags = 520/4 = 130 grams
Rice in each bag = 130 grams

Solution to Question 170

Edward weighs 36 kg.
Edward and Henry together weigh 93 kg.
Henry weighs = 93kg – 36kg = 57 kg
Edward and Paul weigh 87 kg.
Paul weighs = 87kg – 36kg = 51 kg
Total weight of Henry and Paul = 57kg + 51kg = 108 kg

Solution to Question 171

6 kg of flour is enough for making 36 cakes.
With 1 kg of flour, the number of cakes that can be made = 36/6 = 6
Flour needed for 12 cakes = 2 kg

Solution to Question 172

Peter weighs 26 kg.
Melvin is 4 times as heavy as Peter.
Melvin weighs = 26 x 4 = 104 kg
Difference = 104 – 26 = 78 kg
Melvin is 78 kg heavier than Peter.

Solution to Question 173

Two bags of cotton weigh 396 grams altogether.
Bag A + Bag B = 396 grams
Both bags weigh the same after 33 grams of cotton was transferred from Bag A to Bag B.
After transferring, weight of each bag = 396/2 grams = 198grams
Original weight of Bag A = 198grams + 33grams = 231 grams

Solution to Question 174

Two equal pieces of string are tied together.
10 cm from each piece gets used up in tying the knot.
Length of the two strings after tying is 80 cm.
Length of each string after they were tied up = 80/2 = 40 cm
Length of each piece of string before they were tied up = 40 + 10 = 50 cm

Solution to Question 175

There are ten trees in front of John's house.
The distance between the first tree and second tree is 1 m, the second tree and third tree is 2 m, the third tree and fourth tree is 3 m, and so on.

It shows us that there is an increase of 1 m of distance between every consecutive tree.

Distance between the first and the last tree

= 1 + 2 + 3 + 4 + 5 + 6 + 7 + 8 + 9

= 45 m

Solution to Question 176

Claire bought a cloth that was 350 cm long.

She cut it into 5 equal pieces

Each piece's length = 350/5 = 70 centimeters

She used 3 such pieces.

= 3 x 70 = 210 centimeters

Length of the cloth that she had left = 350 – 210 = 140 centimeters

Solution to Question 177

Anne has to travel a distance of 20 km.

She travels 15 km on train and 3.5 km on bus.

She walks the rest of the way.

Distance that Anne had to walk = 20 - 15 - 3.5 = 1.5 km

Solution to Question 178

Dennis traveled from his home to the beach.

He cycled for 3 km and walked 1 km less than the distance he cycled.

Distance Dennis covered by walking = 3 – 1= 2 km

The distance from his house to the beach was 8 km.

He then took a bus for the rest of the journey.

Distance traveled by bus = 8 – 3 – 2 = 3 km

Distance Dennis traveled by bus = 3 kilometers

Solution to Question 179

A ribbon is 18 meters long.

A rope is 125 centimeters long.

1 m = 100 cm
Ribbon = 1800 cm
Difference in length = 1800 – 125 = 1675 meters i.e. 16 meters and 75 centimeters

Solution to Question 180

Pete went swimming.
Each length of the pool was 50 meters long.
He swam 6 lengths = 6 x 50 = 300 meters
Pete needs to swim a total of 500 meters .
Length that remains to be swum = 500 – 300 = 200 meters
Number of lengths Pete is yet to swim so that he has swum 500 meters in total = 200/50 = 4

Solution to Question 181

1 m = 100 cm
10 meters of material = 9 x 100 = 900 centimeters
I need to cut lengths of 30 centimetres = 900/30 = 30
Complete lengths that can be cut = 30
No material will be left over.

Solution to Question 182

Dad needed 7m of wood to build some shelves.
He already had 125 cm of wood.
1 m = 100 cm
7 m = 7 x 100 = 700 cm
Length of wood dad needs to buy = 700 – 125 = 575cm

Solution to Question 183

If a snail travels 3 cm in 5 minutes; distance travelled in half an hour = 3 x 30/5 = 18 cm

Solution to Question **184**

My Car travels 30 km for every one liter of fuel.
A liter of fuel costs $2.
Number of liters of petrol that I can put for $12 = 12/2= 6 liters
Total distance traveled for $12 = 30 x 6 = 180 kilometers

Solution to Question **185**

Containers A, B and C are half-filled with water .
There is 890 ml of water in container A.
120 ml of water in container B and 345 ml of water in container C.
If the 3 containers are completely filled, volume of water in each container will be:
A = 2 x 890 = 1780 ml
B = 2 x 120 = 240 ml
C = 2 x 345 = 690 ml
Total volume that the 3 containers can hold = 1780 + 240 + 690 = 2710 ml

Solution to Question **186**

An empty container was completely filled with water poured from 3 similar jugs.
5 such jugs contained 400 ml of water.
Volume of water in 1 jug = 400/5 = 80 ml
Volume of water in the container = 80 x 3 = 240 ml

Solution to Question **187**

Jeff mixed 840 liters of cold water to 196 liters of lemonade syrup to make
lemonade. Total volume of lemonade = 840 + 196 = 1036 liters
He then poured out 395 liters of the lemonade to serve his guests. Remaining
volume of lemonade = 1036 – 395 = 641 liters
641 liters of lemonade was left.

Solution to Question 188

When a bucket is full, it holds exactly 5 liters of water.
1 liter = 1000 ml
5 x 1000 = 5000 ml
A jug holds 500 milliliters.
Number of full jugs of water needed to fill the bucket = 5000/500 = 10 jugs

Solution to Question 189

The outside of the block has been made with red cubes but the inside part has been made with blue cubes.
A block has 6 faces.
First let us count the total number of cubes in the block = 5 x 4 x 4 = 80
Now count the number of red cubes.
Front face has 5 x 4 = 20 cubes
Back face has 5 x 4 = 20 cubes
The two side faces have 2 x 4 = 8 cubes [cubes along the the edge have already been counted in counting cubes in front and back faces]
Top face has 2 x 3 = 6 cubes [again not counting cubes already counted]
Bottom face has 2 x 3 = 6 cubes [again not counting cubes already counted]
Total number of red cubes = 20 + 20 + 8 + 8 + 6 + 6 = 68 cubes
Number of blue cubes that have been put inside this block = Total number of cubes - number of red cubes = 80 − 68 = 12 cubes

Solution to Question 190

Area of rectangle = L x B = 72 cm^2
a) The table below gives the possible dimensions of the rectangle.

Width	Length
1 cm	72/1 = 72
2 cm	72/2 = 36
3 cm	72/3 = 24
4 cm	72/4 = 18

b) Rectangle that has the largest perimeter = 2 x (L + B)
   72 cm by 1 cm = 2 x (72 + 1) = 146 cm
c) Dimensions of the rectangle with the smallest perimeter = 4cm by 18cm
   Perimeter = 2 x (4 + 18) = 2 x 22 = 44 cm

Solution to Question 191

Area = 48 sq.cm.
Total number of squares in the figure = 12
Area of each square unit = 48/12 = 4 sq.cm
Area of a square = L x B
Therefore side of each square = 2 cm

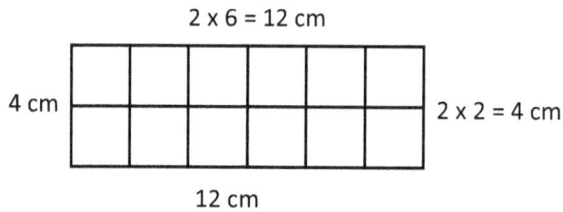

2 x 6 = 12 cm

4 cm

2 x 2 = 4 cm

12 cm

Perimeter of Lisa's figure = 12 + 4 + 12 + 4 = 32 cm

Solution to Question 192

A jogger runs once around a square field.
Each side of the field measures 6500 cm.
Perimeter of a square = 4 x L where L is the length of the square.
100 cm = 1 m
Distance the jogger had run altogether = 4 x 6500 = 26000 cm i.e. 260 m
1 round = 260 m
The total length that the jogger would have run if he goes four and a half times around the field = 260 x 4 + 260/2 = 1040 + 130 = 1170 m
Number of times the jogger would have to run around the field to run a distance of 2600 m = 2600/260 = 10

Solution to Question 193

The length of the field is 3 times its breadth.
Breadth = 24/3 = 8 meters
Perimeter = 24 + 8 + 24 + 8 = 64 meters

It cost $8 per meter to put a fence around the field.
Cost of fencing = 64 x 8 = $512
Cost of fencing the entire field = $512

Solution to Question 194

Breadth of the rectangle = 9 cm
Number of 2 cm lengths that can be cut from breadth = 9/2 = 4 (because we cannot cut half a side)
Length of rectangle = 16 cm
Number of 2 cm lengths that can be cut from breadth = 16/2 = 8
Number of squares that can be cut from the rectangle = 8 x 4 = 32

Solution to Question 195

Perimeter (from point A) going clockwise =
15 + 15 + 15 + 10 + 15 + 10 + 10 + 10 + 10 + 30 = 140 cm

Solution to Question 196

Length of 2 rectangles = 22 − 4 = 18 cm
Length of each rectangle = 18/2 = 9 cm
Area of rectangle = Length x Breadth = 9 x 4 = 36 sq. cm

Solution to Question 197

A wire is cut and bent to form 5 similar rectangles. The length of each rectangle is twice its breadth. The length of a rectangle is 24 cm.
Breadth = 24/2 = 12 cm
Perimeter of the rectangle = 2 x (12 + 24) = 2 x 36 = 72 cm
Perimeter of 5 rectangles = 5 x 72 = 360 cm
The original length of the wire = 3 meter 60 cm

Solution to Question 198

Perimeter of the square = 4 x 15 = 60 centimeters
Perimeter of the rectangle = 2 x (18 + 5) = 2 x 23 = 46 centimeters
Total length = 46 + 60 = 106 centimeters
Half of the piece of wire has been used to form the square and the rectangle
Actual length of the wire = 106 x 2 = 212 centimeters

Solution to Question 199

A room measuring 12 m by 8 m is to be tiled .
Area of the room = 12 x 8 = 96 square meters.
It cost $5 per square meter to tile the room.
Total cost of tiling the room = 96 x 5 = $480

Solution to Question 200

Area of the rectangular plot = 45 x 9 = 405 sq. m
Area of the square = 8 x 8 = 64 sq.m
Area of the swimming pool = 250 sq.m
Area of the land not covered by pool and grass = 405 − 250 − 64 = 91 sq.m

www.ingramcontent.com/pod-product-compliance
Lightning Source LLC
Chambersburg PA
CBHW051343200326
41521CB00014B/2466